WEIRD SPACE SCIENCE

METEOR SHOWERS

THE BREAKDOWN

VIRGINIA LOH-HAGAN

45TH PARALLEL PRESS

Published in the United States of America by Cherry Lake Publishing Group
Ann Arbor, Michigan
www.cherrylakepublishing.com

Photo Credits: © Preston Dyches / Flickr.com via CC BY-NC 2.0 DEED, cover, 1; © ixpert/Shutterstock, 6-7; © Vladi333/Shutterstock, 9; © Benjamin Schaefer/Shutterstock, 11; © Geermy/Shutterstock, 12; © Savvapanf Photo / Shutterstock, 15; © Siberian Art/Shutterstock, 16; © IM_photo/Shutterstock, 19; © Nuttapong/Shutterstock, 20

Graphic Element Credits: Cover, multiple interior pages: © WindAwake/Shutterstock, © kichikimi/Shutterstock, © Your Local Llamacorn/Shutterstock, © Marishiav/Shutterstock, © Supza/Shutterstock, © HSSstudio/Shutterstock, © Olive Kitt/Shutterstock, © KKarance/Shutterstock, © Theo_hrm/Shutterstock, © Rodin Anton/Shutterstock, © Petr Vaclavek/Shutterstock, © Nikolaeva/Shutterstock, © Evgenii Doljenkov/Shutterstock

45th Parallel Press is an imprint of Cherry Lake Publishing Group.

Library of Congress Cataloging-in-Publication Data has been filed and is available at catalog.loc.gov

Cherry Lake Publishing Group would like to acknowledge the work of the Partnership for 21st Century Learning, a Network of Battelle for Kids. Please visite Battelle for Kids online for more information.

Printed in the United States of America

Dr. Virginia Loh-Hagan is an author, university professor, and former classroom teacher. She's currently the Director of the Asian Pacific Islander Desi American Resource Center at San Diego State University. She loves science-fiction stories. She lives in San Diego with her very tall husband and very naughty dogs.

TABLE OF CONTENTS

BREAK IT DOWN

1. DECODE IT

- Look at each letter.
- Look at letter patterns.
- Blend sounds.
- Try other possible sounds.

2. BUILD VOCABULARY

- Listen to the word.
- Match the sounds to a word you know.
- Look for context clues for words you don't know.
- Look up the meaning in a glossary or dictionary.

3. MAKE CONNECTIONS

- Name other examples.
- Draw the idea.
- Compare ideas.
- Ask questions.

TEST IT OUT

WORDS YOU'LL LEARN

Look over the list of vocabulary words. Are there any you already know?

atmosphere
constellations
friction
meteorites
meteors
meteor showers

GET STARTED

Open and closed syllables are 2 types of syllables. Open syllables end with a long vowel. Closed syllables end with a consonant. They have a short vowel sound.

- Look at the letters.
- Divide words into syllables.
- Try dividing another way if needed.

The word *friction* starts with a closed syllable. It is pronounced: FRIK-shuhn.

- the 1st vowel is a short *i*: *frik-*.
- The 2nd vowel says "uh": *-shuhn*.

Planet Earth has an **atmosphere**.

This is a layer of gasses. It surrounds a planet or other object.

Sometimes space rocks enter Earth's atmosphere. They're called **meteors**.

When a meteor falls to Earth, it drags as it falls.

MAKE CONNECTIONS

Friction is when motion is resisted. It makes heat.

- Skate fast across a carpet in socks. Feel the bottom of your socks. Are they cold? Are they warm?
- How might friction affect meteors?

The air in the atmosphere pushes against the meteor.

Meteors burn brightly as they fall to Earth. It looks like a bright streak.

The bright streak is actually glowing hot air. It's not the rock.

Sometimes, lots of meteors fall to Earth.

They fall at the same angle.

This is called a **meteor shower**.

DECODE IT

The word *meteor* has 2 open syllables.

- Break it into chunks: me-te-or.
- The 1st and 2nd vowels are a long *e*.
- The 3rd vowel sound says "uh."
- *Meteor* is pronounced MEE-tee-uhr.

Meteor showers are named for **constellations**.

BUILD VOCABULARY

A **constellation** (kahn-stuh-LAY-shuhn) is a group of stars.

- It makes a picture in the sky.
- Ursa Major is a constellation.
- Other forms: constellations, constellate, constellational

Constellations are groups of stars.

The Big Dipper, or Ursa Major, can be seen from the Northern Hemisphere.

Some meteors are big rocks.

But most are small.

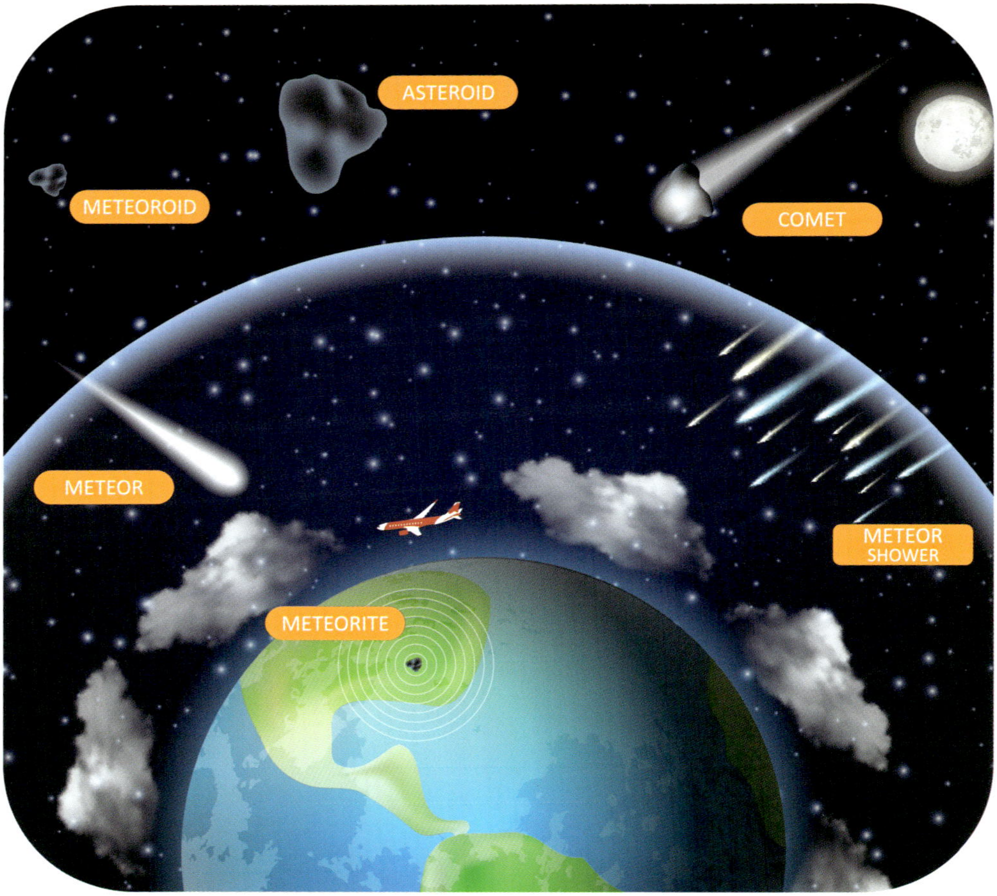

They're often smaller than a grain of sand.

Most meteors burn up in the atmosphere.

This means they won't hit Earth's surface.

There's no harm to humans.

Sometimes meteors survive through Earth's atmosphere.

Bits may fall to Earth.

They're known as **meteorites**.

DECODE IT

The word *meteorites* has 4 vowel sounds.

- Break it into chunks: me-te-o-rites.
- The 1st and 2nd vowel sounds are each a long *e*.
- The 3rd vowel sound says "uh."
- The last vowel sound is a long *i*.
- *Meteorites* is pronounced MEE-tee-uh-riyts.

PUT IT TOGETHER

Main Idea:
Meteors are space rocks that enter Earth's atmosphere.

Detail:
Meteors that fall to Earth are called meteorites.

Detail:
An atmosphere is a layer of gases that surrounds a planet. Earth has one.

Detail:
Meteor names come from constellations.

Detail:
Meteors make a bright streak of light in the sky. This is caused by friction.

Detail:
When lots of meteors fall at the same time it's called a meteor shower.

LEARN MORE

BOOKS

Loh-Hagan, Virginia. *Introduction to Space*. Ann Arbor, MI: 45th Parallel Press, 2025.

Loh-Hagan, Virginia. *Weird Science: Space*. Ann Arbor, MI: 45th Parallel Press, 2022.

Tyson, Neil deGrasse. *StarTalk: Young Readers Edition*. Washington, DC: National Geographic Kids, 2018

ONLINE

Search these online sources with an adult:

- NASA | Games and Interactives
- NASA+ | Explore Topics

GLOSSARY

atmosphere (AT-muhs-feer) gases surrounding Earth or other objects in space

constellations (kahn-stuh-LAY-shuhnz)groups of stars

friction (FRIK-shuhn) when motion is resisted; creates light and heat

meteorites (MEE-tee-uh-riytz) bits of meteors that make it to Earth's surface

meteors (MEE-tee-uhrz) space rocks that enter Earth's atmosphere

meteor showers (MEE-tee-uhr SHOW-uhrz) a phenomenon when lots of meteors fall to Earth at the same time, from the same angle

INDEX